AF374801

Fundamental Constants of Cosmos Theory and The Standard Model

Dimensions Related to Interactions
Coupling Constants' Sequence of Powers of 2
Coupling Constant Values are Measures of Degrees of Freedom
Cosmos Parent Universe Consistency Condition
Fundamental Constants as Simple Functions of e and pi
Weinberg Angle Calculation
Reduction of Everything to Degrees of Freedom

Stephen Blaha Ph. D.
Blaha Research

Pingree-Hill Publishing
MMXXIV

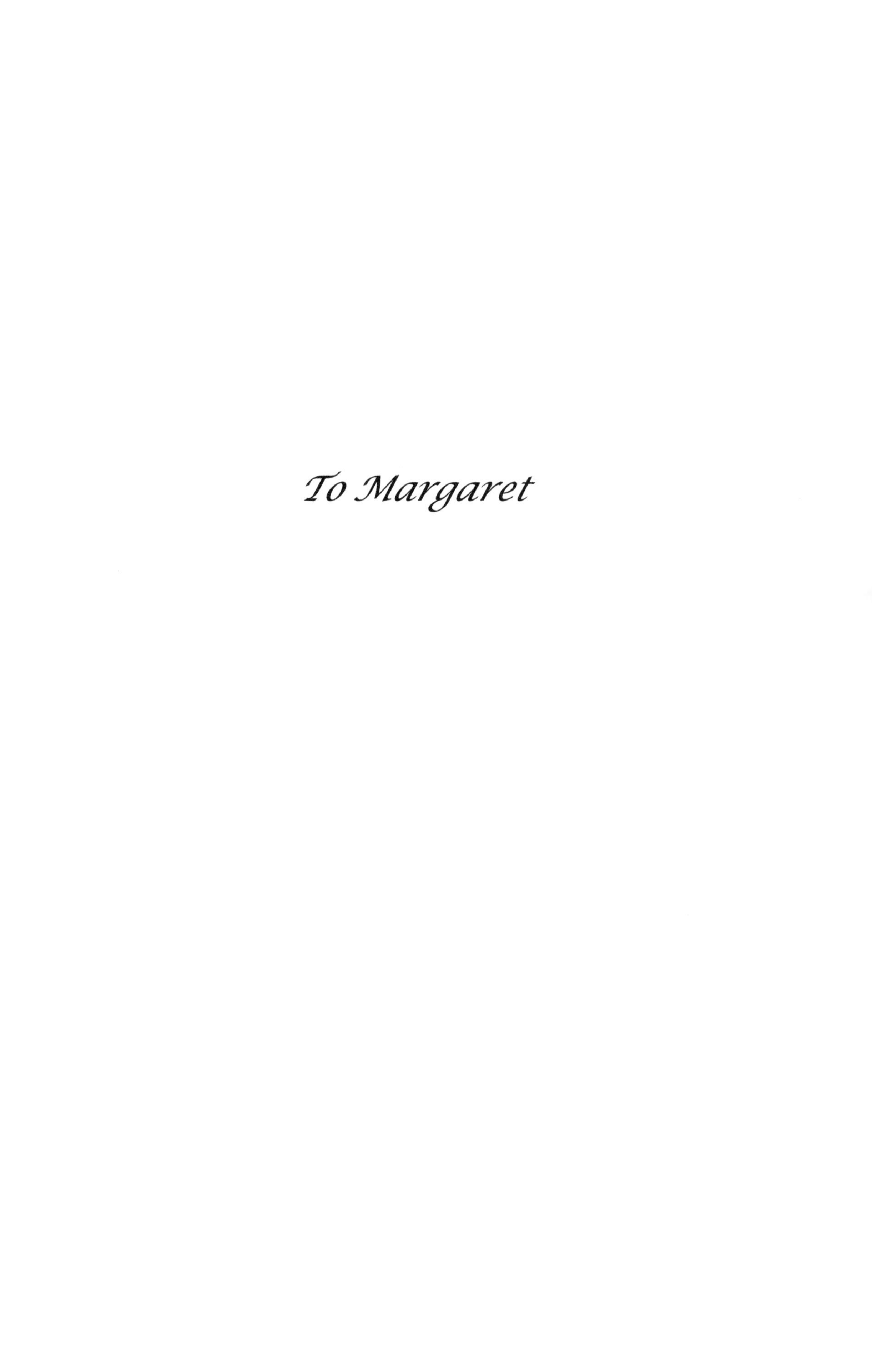

To Margaret

Some Other Books by Stephen Blaha

SuperCivilizations: Civilizations as Superorganisms (McMann-Fisher Publishing, Auburn, NH, 2010)

All the Universe! Faster Than Light Tachyon Quark Starships & Particle Accelerators with the LHC as a Prototype Starship Drive Scientific Edition (Pingree-Hill Publishing, Auburn, NH, 2011).

Unification of God Theory and Unified SuperStandard Model THIRD EDITION (Pingree Hill Publishing, Auburn, NH, 2018).

The Exact QED Calculation of the Fine Structure Constant Implies ALL 4D Universes have the Same Physics/Life Prospects (Pingree Hill Publishing, Auburn, NH, 2019).

Passing Through Nature to Eternity ProtoCosmos, HyperCosmos, Unified SuperStandard Theory (Pingree Hill Publishing, Auburn, NH, 2022).

HyperCosmos Fractionation and Fundamental Reference Frame Based Unification: Particle Inner Space Basis of Parton and Dual Resonance Models (Pingree Hill Publishing, Auburn, NH, 2022).

The Cosmic Panorama: ProtoCosmos, HyperCosmos, Unified SuperStandard Theory (UST) Derivation (Pingree Hill Publishing, Auburn, NH, 2022).
Ultimate Origin: ProtoCosmos and HyperCosmos (Pingree Hill Publishing, Auburn, NH, 2022).

God and and Cosmos Theory (Pingree Hill Publishing, Auburn, NH, 2023).

Newton's Apple is Now The Fermion (Pingree Hill Publishing, Auburn, NH, 2023).

Cosmos Theory: The Sub-Particle Gambol Model (Pingree Hill Publishing, Auburn, NH, 2023).

Cosmos-Universe-Particle-Gambol Theory (Pingree Hill Publishing, Auburn, NH, 2024).

Fractal Cosmos Curve: Tensor-based Cosmos Theory (Pingree Hill Publishing, Auburn, NH, 2024).

The Eternal Form of Cosmos Theory Third Edition (Pingree Hill Publishing, Auburn, NH, 2024).

Available on Amazon.com, bn.com Amazon.co.uk and other international web sites as well as at better bookstores.

CONTENTS

FIGURES and TABLES

Introduction

The growing interest in Cosmos Theory, and some major new significant results, have motivated this new book. It covers a host of topics beginning with a more detailed derivation of the Consistency Condition for the origin of the Cosmos that was presented in earlier books.

From there it proceeds to describe new results that stem, in part, from the consideration of the Consistency Condition. The consequence of the book: There is a unifying principle based on degrees of freedom, which is at the basis of Cosmos Theory and is found in its spaces, symmetries, universes, universe structure, energies, and particle interactions.

The book covers numerous topics including:

1. The $r = 18$ space is shown to be the first even dimension integer (most minimal) space whose universe does not collapse. Thus it is an optimal choice for the parent Cosmos space for an $r = 18$ universe. Within that universe sub-universes may be created that may undergo Hubble expansion. It is possible that there are many such unrelated parent universes with their own sub-universes.

2. Due to the factor $(2e\pi/r)^r$ the dimension $r = 17.08$ marks the critical dimension for the transition from a contracting universe.

3. The consistency condition embodies a factor $e\pi/4 = 2.13$ where $e = 2.718$ is the natural logarithm base constant. It appears in the factor $(2e\pi/r)^r \cong (17.08/r)^r$ where r is the dimension. Dimension thus has a numerical origin.

4. This author proposes a new constant of Nature c_g, which appears in many contexts in Astrophysics and Elementary Particle Physics. $c_g = \pi^2/128$. The relation between temperature and mass of an entity is given by $kT = mc_g$.

5. The book shows a variety of fundamental constants may be expressed in terms of simple consistent expressions using the natural logarithm base e and $\pi = 3.14159$.

6. Coupling constants values for ElectroWeak and Strong coupling constants are shown to exhibit a sequence in powers of 2 just like Cosmos Theory dimension arrays.

7. The Weinberg angle.is calculated to within 8%.

8. Coupling constants are shown to be equivalent to products of factors wherein each factor represents degrees of freedom.

Everything is a manifestation of degrees of freedom: dimensions, internal symmetry coupling constants, dimension arrays, universes, and spaces.

1. Origin of the Tenth Cosmos Space and Universe

We begin this book with a major extension of the consideration of the generation of the initial Parent universe. We will see that it leads to a set of expressions for important constants such as the origin of the power $2 \cong e\pi/4$; the choice of 10 Cosmos spaces; the choice of 18 space-time dimensions; the choice of $c_g = 0.0785 \cong \pi^2/128$ which appears in a universal relation $kT = mc_g$, as a new universal constant. Remarkably, in addition, we find the Standard Model interaction coupling constants also have a powers sequence in 2 clearly suggesting a geometric origin of the Standard Model interactions. The values of the coupling constants, and other quantities are directly expressible in fundamental constants, π and e (the base of Natural logarithms), and simple integer powers of 2. The values of coupling constants seem to depend on fermion nature (geometric) and geometry.

The conclusion of this work is:

Geometry is the Basis of Cosmos Theory,
the Standard Model, and Internal symmetry Interactions.

1.1 The Origin of the Cosmos

We initially chose an $r = 18$ space as the space of the original Parent universe raises several issues. (We now prove it.)

The logic of the origin point: There can be no time before the definition of the Parent space dimension array since no space or universe existed before its definition. Yet a Parent universe cannot exist without the definition of the Parent's dimension array. Thus the dimension array and Parent universe must be simultaneously defined and created.

The only way that one may reasonably resolve this state of affairs is to define both as the consequence of a consistency condition. We suggest the consistency condition for the minimum Parent universe dimension is the requirement that its virtual thermodynamic outward pressure equals the Casimir vacuum inward pressure from its virtual external vacuum Casimir pressure at the point of origin of the virtual universe.[1] See Fig. 1.1. Below this minimum dimension the Parent universe cannot exist since the virtual vacuum pressure would cause it to be contracted into non-existence.

The dimension r of the Parent universe fixes a maximum Physical Cosmos space dimension.[2] Within the Parent universe there are universes of lower dimension corresponding to lower dimension Cosmos spaces.

---

[1] Much of this chapter is based on Blaha (2022d) and (2023e).
[2] Depending on whether the Parent universe is mildly-expanding ($r = 18$) or expanding ($r = 20$). We discuss this issue later.

1.2 Balance Between Expansion and Contraction

When a possible virtual universe is "virtually" created it may expand (Hubble expansion) or it may contract to a point and then cease to exist, or it might be static. The decision is based on the dynamics of the virtual creation. Does the virtual universe possess sufficient mass-energy to generate a pressure for expansion or does the potential surrounding[3] virtual vacuum pressure causes the universe to contract and be "snuffed out"?

The balance between inward and outward pressure can be represented with

$$H = P - C \tag{1.1}$$

where P is the pressure of the energy-momentum contents of the virtual universe and C is the Casimir force due to the virtual vacuum energy embodied in the surface of the universe. There are three possibilities:

$$
\begin{aligned}
&H < 0 \qquad \text{Casimir force causes contraction to a point} \\
&H = 0 \qquad \text{Balance of forces – "stability"} \\
&H > 0 \qquad \text{Pressure causes (Hubble) expansion of universe}
\end{aligned}
\tag{1.2}
$$

We wish to use H to determine the Cosmos Theory Physical spectrum. It has a lower limit at $r = 0$ but no apparent upper limit. We now view H as determining the maximum dimension r_{Parent} of the Physical Cosmos spectrum:

$$
\begin{aligned}
&H \geq 0 \qquad \text{"Normal" universe } r \leq r_{Parent} \\
&H < 0 \qquad \text{Virtual universe lapses into non-existence} \\
&H = 0 \qquad \text{Static universe}
\end{aligned}
\tag{1.3}
$$

We now proceed to determine H for the space of dimension r.

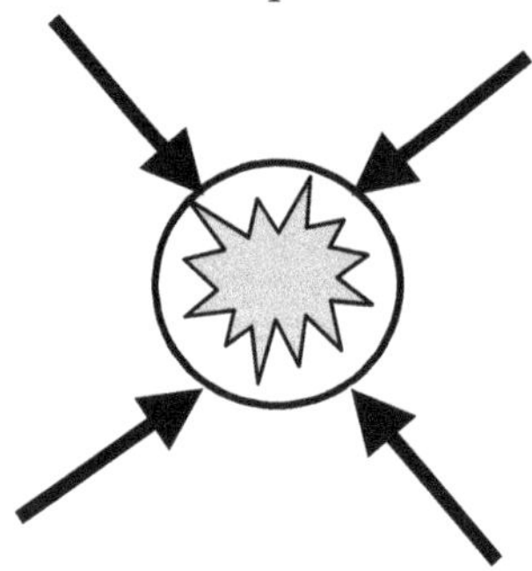

Figure 1.1. Schematic diagram of virtual universe's "vacuum" bubble showing the confining Casimir force of the exterior vacuum. This force is countered by internal fermion gas pressure symbolized by a jagged interior star.

---

[3] In the *surface* of the virtual universe.

1.3 Number of Fundamental Particle Species

Cosmos Theory specifies the number of fundamental fermion species N for a universe in a space of dimension r:

$$N = 2^{r+4} \qquad (1.4)$$

The determination of the maximum number of fermions from H will fix the maximum dimension r.

1.4 Conditions at the Consistency Point

We now consider an infinitesimal box containing the energy-momentum of a virtual universe. The box contains fermions with the initial features:

1. All fermions are massless.
2. All interactions are zero.
3. All baryon numbers are zero.

1.5 Thermodynamic Outward Pressure

We base the thermodynamic pressure on the Ideal Gas Law[4]

$$P = NkT/V \qquad (1.5)$$

where P is the pressure, T is the temperature, k is Boltzmann's constant, and N is the number of particles in unit volume V.

Assuming the fermion gas occupies a spherical spatial ball with spatial volume

$$V_{r-1} = \pi^{(r-1)/2} a^{(r-1)}/\Gamma((r-1)/2 + 1) \qquad (1.6)$$

with radius a in an r dimension space-time:

$$P = N \, \Gamma((r-1)/2 + 1)kT/(\pi^{(r-1)/2} a^{(r-1)}) \qquad (1.7)$$

where $\Gamma(n)$ is the factorial function. The number of fermions N is given by eq. 1.4.

1.6 Casimir Effect Confining Force

This section calculates the force on a universe due to the Casimir effect.[5] We assume the universe is in a vacuum energy bubble in an r − 1 dimension spatial region of radius a. It generates a confining force on the surface of the universe. We start with the infinite Dirac fermion vacuum energy per unit volume:

---

[4] One might think of introducing more complex forms of gas laws based on the volume occupied by the gas particles and so on. However our assumption of zero interactions at the origin of universes eliminates the need for such "refinements." One might also thinking of introducing an attractive gravitation term in the consistency condition. Here again the assumption of no interactions at the origin point removes that concern.

[5] H. G. B. Casimir, Koninkl. Ned. Adak. Wetenschap. Proc. **51**, 793 (1948), S. K. Lamoreaux, Phys. Rev. Lett. **78**, 5 (1997), L. S. Brown and G. J. McClay, Phys. Rev. **184**, 1272 (1969), and references therein.

$$U_\infty = \int_0^\infty d^{r-1}k \, (k^2 + m^2)^{\frac{1}{2}} \tag{1.8}$$

where m is the common fermion mass, which we take to be zero by section 1.4. Due to the vacuum-less bubble, the vacuum energy changes to

$$U = U_\infty - U_{bubble} \tag{1.9}$$

where the missing bubble vacuum energy per unit volume is

$$U_{bubble} = \int_d^\infty d^{r-1}k \, (k^2 + m^2)^{\frac{1}{2}} \tag{1.10}$$

where d is the momentum space lower limit of the excluded vacuum energy within the bubble. The Casimir deficit is

$$U = \int_0^d d^{r-1}k \, k \tag{1.11}$$

Assuming the momentum d corresponds to the spatial radius, a, of the bubble

$$a = 1/(mc) = 1/d \tag{1.11a}$$

where c is the speed of light we find the energy deficit per unit volume is

$$U \cong \pi^{(r-1)/2} a^{-r} \, (r-1)/[r\Gamma((r-1)/2 + 1)] \tag{1.12}$$

Multiplying by the volume of the bubble we find a total bubble energy:

$$U_{tot} \approx V_{r-1} \, U \tag{1.13}$$

$$\approx \pi^{r-1} a^{-1}[(r-1)/r]/\Gamma((r-1)/2 + 1)^2 \tag{1.14}$$

Then

$$\partial U_{tot}/\partial a \cong -\pi^{r-1} a^{-2}[(r-1)/r]/\Gamma((r-1)/2 + 1)^2 \tag{1.15}$$

A bubble surface of radius a has the area

$$S_{r-2} = 2\pi^{(r-1)/2} \, a^{r-2}/\Gamma((r-1)/2) \tag{1.16}$$

The Casimir force on the surface is

$$C = (S_{r-2})^{-1} \, \partial U_{tot}/\partial a \tag{1.17}$$

$$= -\tfrac{1}{2}\,\pi^{(r-1)/2}\,a^{-r}\,[(\,r-1)/r]\,\Gamma((r-1)/2)/\Gamma((r-1)/2+1)^2$$
$$= -\,\pi^{(r-1)/2}\,a^{-r}/[r\Gamma((r-1)/2+1)]$$

1.7 H Balance

We now set the outward pressure of the internal fermion "gas" equal to the inward Casimir vacuum force by eqs. 1.7 and 1.17:

$$H = P - C$$
$$= N\,\Gamma((r-1)/2+1)kT/(\,\pi^{(r-1)/2}a^{(r-1)}) - \pi^{(r-1)/2}\,a^{-r}/[r\Gamma((r-1)/2+1)] \qquad (1.18)$$

Eliminating common factors in H we find the condition becomes

$$H' = N\,akT - \pi^{(r-1)}/[r\Gamma((r-1)/2+1)^2] \qquad (1.19)$$

At the critical point of eq. 1.2 we see $H' = 0$ implying

$$N = \pi^{(r-1)}/[arkT\,\Gamma((r-1)/2+1)^2] \qquad (1.20)$$

For collapse $H' < 0$
$$N < \pi^{(r-1)}/[arkT\,\Gamma((r-1)/2+1)^2] \qquad (1.21)$$

For expansion $H' > 0$
$$N > \pi^{(r-1)}/[arkT\,\Gamma((r-1)/2+1)^2] \qquad (1.22)$$

1.8 N and H' as Functions of Dimension r

We find that the choice of

$$akT = 2.635 \times 10^{-10} \qquad (1.23)$$

and the Cosmos dimension array size choice $N = 2^{r+4}$, give a critical point at dimension $r = 18$.[6] Thus at minimum an $r = 18$ Parent universe can exist.

For even number dimensions below $r = 18$ (assuming eq. 1.23) we find H' is negative indicating such a Parent universe could not exist. The Casimir force prevents its existence. At $r = 18$ the pressure is equal to the Casimir force thus allowing an 18 dimension Parent universe within which child universes may be created. The $r = 18$ virtual universe is the minimal[7] (in r) universe, and thus the Parent universe.

On this basis we can choose a Physical set of ten Cosmos spaces. We can create one or more 18 dimension Parent universes. Each Parent universe is stationary. It balances expansion and contraction. A parent universe can contain universes of lesser Cosmos dimension since they have a less strong Casimir force with which to contend.

---

[6] Later in section 1.10 we refine the choice of critical dimension to $r = 17.08$ based on the critical turning point of the factor $(2e\pi/r)^r$ at $r = 17.08$ of $2e\pi/r$ from greater than 1 (contraction) to less than 1. A minimum $r > 17.08$ is required to avoid collapse to a point. We chose the next greater even integer $r = 18$ as the Parent space dimension.

[7] We require the dimension to be an even number.

If we wish the Parent universe to be more rapidly *expanding* then we can choose the r = 20 space's universe as the Parent universe. This choice would expand the Physical set of HyperCosmos spaces to eleven.

1.9 Support for the Power Series Form of Dimension Arrays

The size of dimension arrays N is a power series in 2 as shown in eq. 1.4. If we use eq. 1.20, and use Stirling's approximation for the gamma function, a power series emerges with exponent r for large values of r.

Stirling's approximation is

$$n! = (2\pi n)^{\frac{1}{2}} (n/e)^n \qquad (1.24)$$

For large r >> 1, n ≅ r/2 and

$$\Gamma((r-1)/2 + 1) \rightarrow (\pi r)^{\frac{1}{2}}(r/2e)^{r/2}$$

Substituting in eq. 1.20 gives

$$N = \pi^{(r-1)}/(arkT) \times 1/(\pi r(r/2e)^r)$$
$$= (2e\pi/r)^r/(\pi arkT) \qquad (1.25)$$
$$= (17.08/r)^r/(\pi arkT)$$

It shows the same form of power series with exponent r as N in eq. 1.4. The similarity in form, which is based on the form of volume and surface area in higher dimension spaces, reflects their joint basis in differential forms, n-forms and wedge expressions, seen before.

We see a strengthening of the view of dimension array sizes as based on geometry.

1.10 Why Ten Physical HyperCosmos Spaces?

An examination of eq. 1.25 shows that there is a more refined critical point value due to the factor

$$(2e\pi/r)^r \qquad (1.26)$$

where the number $2e\pi/r$ transitions from greater than one to less than one at r = 2eπ = 17.08. At r = 18 the Casimir force has changed from greater than the thermodynamic pressure to less than the thermodynamic pressure OR from contraction to expansion of the Parent universe. Thus r = 18 marks the beginning of an acceptable expanding Parent universe. At r < 18 universes contract to a point and thus could not generate a Parent universes. Parent universes can have expanding child universes within them.

The key scale for universe, and thus space dimension, is 2eπ = 8(eπ/4) ≅ 18. We view this result as a confirmation of the approach in this chapter and the choice of r = 18 as the largest possible Physical HyperCosmos space as well as of the 10 Physical HyperCosmos space spectrum.

1.11 Implications of Eq. 1.25

The consistency condition that we have found has a number of important results including the possibility of a new constant of nature c_g = 0.0785. This constant appears in many contexts in Elementary Particle theory and Astrophysics. See chapter 2.

The implications of the consistency condition are:

1. The $r = 18$ space is the first even dimension integer (most minimal) space whose universe does not collapse. Thus it is an optimal choice for the parent space for an $r = 18$ universe. Within that universe sub-universes may be created that may undergo Hubble expansion. It is possible that there are many such unrelated parent universes with their own sub-universes.

2. Due to the factor $(2e\pi/r)^r$ the number 17.08 marks a critical dimension for the transition from a contracting universe.

3. The consistency condition embodies a factor $e\pi/4 = 2.13$ where $e = 2.718$ is the natural logarithm base constant. It appears in the factor

$$(2e\pi/r)^r = (8\times(e\pi/4)/r)^r = (8\times(e\pi/4)/r)^r \cong (17.08/r)^r \qquad (1.27)$$

4. Thus $e\pi/4 \cong 2$ is the constant in the dimension array d_{dr} series of powers to within 7% - a good result considering the approximate nature of the consistency condition calculation.

5. The value of the eq. 1.25 factor

$$akT$$

with the consistency condition value (eq. 1.23)

$$akT = 2.635 \times 10^{-10} \qquad (1.28)$$

has an important further interpretation when kT is approximated by[8]

$$kT = mc_g \qquad (1.29)$$

where

$$c_g = 0.0785 \qquad (1.30)$$

Using $a = 1/mc$ (where c is the speed of light $c = 3 \times 10^8$ m/s from eq. 1.11a) we find

$$akT = 0.0785/c = 2.62 \times 10^{-10} \qquad (1.31)$$

which differs by 2.4% from the consistency condition value above.

---

[8] The constant c_g appears in many contexts: elementary particles, universe studies, and Cosmos universes studies. See the next section for details.

6. When the $kT = mc_g$ relation is substituted in eq. 1.25 we find the consistency condition is independent of the mass m.

2. A New Constant of Nature c_g

The author's new constant c_g appears in many contexts in our studies in earlier books:

Blaha (2023e) *Cosmos Theory: The Sub-Particle Gambol Model*
Describes its use in deep inelastic lepton and quark resonances, production and interactions, hadron-hadron collisions; neutrino oscillations e - μ, $\mu - \tau$; particle stability pressure vs. Casimir force, and Hubble expansion. See section 9.8 and 9.9. In section 9.9 (eq. 13.22) we find an approximate relation, which we now express as

$$c_g = \pi^2/128 = 0.0771$$

giving a 2% difference from the approximate relation we found $c_g = 0.0785$.

Blaha(2024a) *Cosmos-Universe-Particle-Gambol Theory*
Here we found the following uses of c_g:

> The Cosmos Theory Distribution of Universes
> Anti-Planckian Model of Hubble Expansion Parameter
> Our Universe as a Megaverse Particle
> Universe-Megaverse Boundary
> Casimir Force Limit on Hubble Expansion
> S_8 Universe-Gambol Clumping Models
> Variation of Universe Cluster Mass-Energy Sizes
> Neutron Star Quark Core Gambol Model
> GVDM Photon $- \rho$ meson Gambol Model
> Higgs and Massive Vector Boson Gambol Models

The close matches in values of $akT = 2.635 \times 10^{-10}$, of $kT = mc_g$, and of c_g values seen in chapter 1 suggest it is a new universal constant.

3. A New Level of Physical Reality

The calculation of the consistency condition for Cosmos Theory has led to a number of simple expressions based on fundamental mathematical constants that raise the possibility of a new deeper level of Reality at the base of Cosmos Theory and Physics in general:

Quantity	Expression	Value	Known Value	Deviation
d_{dn} Power series Variable	$e\pi/4$	2.13	2	6.5%
$c_g = kT/m$ constant	$\pi^2/128$	0.0771	0.0785	1.8%
Consistency Condition dimension	$8\,e\pi/4$	17.08		
akT	c_g/c	2.62×10^{-10}	2.635×10^{-10}	2.4%

To these values may be added a similar looking expression for α:

Fine Structure Constant[9] α	$e^2/1024$	0.00721438	0.0072973525643	1.15%

where e is the natural logarithm base $e = 2.718$. The approximation compares extremely well with the experimentally known value of α. The Fine structure constant appears as a constant in QED and ElectroWeak theory. *Its appearance above with an analogous form of expression suggests that it has a geometric origin.* The constant e that appears in the above calculations follows from Stirling's approximation for gamma functions appearing the expressions for higher dimension volumes and surface areas.

The relative closeness of the calculated value of "fine structure constant" to the experimentally known value is very encouraging.

The expression for the Fine Structure Constant above seems to be repeated for the other Standard Model interactions' coupling constants. We note factors of four between the known values suggesting a correspondence to the factors of four in Cosmos Theory. The calculated and experimental values below are based on powers of two is close to the known experimental values.[10]

---

[9] Calculated to known experimental value in the Johnson-Baker-Willey QED by this author. See Blaha (2019f) and references therein.

[10] The appearance of SU(4) is speculated. The interactions for SU(n) interactions for $n > 4$ are presently unknown. If such interactions exist then their strength might lead to total confinement such as we see for quarks.

	ESTIMATE			**EXPERIMENT**		
Interaction	**Expression** $g^2/4\pi$	**Value**	**g**	**Known Value** $g^2/4\pi$		**Deviation**
U(0)	$e^2/4096$	0.0018036	-	-		-
U(1) α	$e^2/1024$	0.00721438	0.303	0.0072973525643		1.15%
SU(2) g	$e^2/256$	0.0289	0.63	0.0316		9.3%
SU(3) α_S	$e^2/64$	0.115	1.21	0.117		1.7%
SU(4)[11] ?	$e^2/16$	0.462	2.4?	0.458		0.087%

where g is the known coupling constant value[12] and e = 2.718 is the natural logarithm base. *Internal symmetry coupling constants have geometric values*[13].

There appears to be a sequence of powers in the expressions for the coupling constants:

$$SU(n) \qquad\qquad g(n)^2/4\pi = e^2/2^{12-2n} = 2^{r-12}\,e^2$$

$$g(n) = e\,(4\pi)^{\frac{1}{2}}\,2^{r/2-6} = 9.635\,\,2^{r/2-6} = 0.1505\,\,2^{r/2} \qquad\qquad (3.1)$$

where r = 2n is the number of *real* dimensions in the group's fundamental representation. The decrease in $g^2/4\pi$ by factors of four is analogous to the quadrupling effect in dimension array size in Cosmos Theory. *Thus interactions, and their strengths, appear to be geometric in origin.*

3.1 Weinberg Angle θ_W Calculation

The Weinberg angle is related to the electromagnetic and Weak coupling constants with

$$e^2/g^2 = \sin^2\theta_W$$

where e is the electric charge. From the above table, using our calculated values, we find

$$\sin^2\theta_W = 256/1024 = 0.25$$
$$\sin\theta_W = 0.5$$

compared to the experimental value of $\sin^2\theta_W = 0.231$. An accuracy to 8% that further supports the power sequence of eq. 3.1.

Since the Weinberg angle is related to massive vector boson masses we see the conventional definition gives

$$m_W/m_Z = \cos\theta_W = 0.866$$

---

[11] This value is based on the "doubling trend" seen in the three known coupling constants above.
[12] All coupling constant values are based on data from Particle Data Group 2022.
[13] We add a U(0) line for the sake of completeness.

in our approximation.

The Coupling Constants, the dimension arrays, and the relation of coupling constants to masses (and therby to energies) all reflect their common interpretation as manifestations of degrees of freedom. There is a unifying principle at the basis of Cosmos Theory in its spaces, symmetries, universes, universe structure, energies, and particle interactions.

4. Coupling Constants as Dimensions

Coupling constants (excepting gravitation) are numbers. We have shown that they have the same sequence of powers as we saw in dimension arrays.

We can treat the dimension array of dimension r with the size:

$$d_{dr} = 2^{r+4}$$

as corresponding to a special unitary group, which is the largest possible group in dimension r,

$$MAX = SU(2^{r+3})$$

based on the number of dimensions in the group's fundamental representation.

From the previous expression for $g^2/4\pi$, the dimension array size in ratio to the coupling constant is

$$(g_{max}(n)^2/4\pi)/d_{dr} = 2^{r-12} e^2/2^{r+4} = 2^{-16}e^2$$

in dimension r where $n = d_{dr}/2$. Thus d_{dr} dimensions are required for $SU(2^{r+3})$ and

$$d_{dr} = 2^{16}g_{max}^2/(4\pi e^2) \tag{4.1}$$

One dimension has the coupling constant value

$$2^{12-r}g_{max}^2/(4\pi e^2) \tag{4.2}$$

We now have a direct relation in dimension r between the dimensions in a dimension array and the size and coupling constant of the maximal group in dimension r.

4.1 Dimension Cost of an Interaction Coupling Constant

Each interaction has an associated coupling constant, which is related to the group of the interaction:

$$SU(n) \qquad\qquad g(n)^2/4\pi = e^2/2^{12-2n}$$

If we ask what the "cost" is, in dimensions, of a coupling constant we find

$$\text{Cost in dimensions} = 2^{2n-12} e^2/2n \ \text{ per dimension}$$

4.2 Coupling Constants as Products of Dimension Degrees of Freedom

We have shown coupling constants appear to form a powers sequence as seen in eq. 3.1. We now consider the expression of a coupling constant as the product of a number of factors that are based on geometry.

First we note that n is the number of fermions in a fundamental representation of SU(n). The coupling constant g has the form:

$$g(n) = e\,(4\pi)^{\frac{1}{2}}\,2^{n-6} = e\,(2^{-10}\pi)^{\frac{1}{2}}\,2^n = 0.1505\;2^n \tag{4.3}$$
$$= g_g\,(\text{number of spin states per fermion})^{\text{number of fermions}}$$

The coupling constant is a fundamental value times the number of spin states (a geometric property of fermions equal to half the column length of a spinor) raised to the number of fermions.

We see that g(n) has n factors – one factor for 2 spin states for each fermion in the SU(n) fundamental representation. The origin of

$$g_g = e\,(2^{-10}\pi)^{\frac{1}{2}} \tag{4.4}$$

is not fully understood due to the appearance of the natural logarithm base e.

Eq. 4.3 implies that the coupling g(n) for SU(n) is g_g times n factors of 2 which represents the number of spin states per fermion in the SU(n) fundamental representation in four dimensions. Thus it is a product of degrees of freedom.

An important point is the number of factors is equal to the number of fermions in the SU(n) fundamental representation. *The coupling constant "felt" by one fermion depends on the total number of fermions in the group's fundamental representation.*

4.3 A Possible Generalization to Higher Space-time Dimensions

Note g_g has the form

$$g_g = e\,(2^{-12}\,S_2)^{\frac{1}{2}} \tag{4.5}$$

where S_2 is the surface area of a unit 3-sphere in four dimensions. (eq. 1.16)

We can generalize eq. 4.3 to higher dimensions using the column length of higher space-time dimensions r:

$$g^r(n) = e\,(2^{-12}S_{r-2})^{\frac{1}{2}}\,(2^{r/4})^n = e\,(2^{-12}S_{r-2})^{\frac{1}{2}}\,2^{nr/4} \tag{4.6}$$
$$= g^r{}_g(n)\,2^{nr/4}$$

using eq. 1.16 for the surface of a unit (r – 1)–sphere in r dimensions.

The constants $g^r{}_g(n)$ have the form

$$g^r{}_g(n) = e\,(2^{-12}\,S_{r-2})^{\frac{1}{2}} \tag{4.7}$$

They are geometric. The factor of e is discussed in the following section.

4.4 Form of $g^r_g(n)$

We can expand

$$g^r_g(n) = e \, (2^{-12} \, S_{r-2})^{\frac{1}{2}}$$

using the surface area S_{r-2} of eq. 1.16 and Stirling's approximation for $\Gamma(n)$ in eq. 1.24.

We find the approximate expression in terms of e and π:

$$g^r_g(n) \sim e^{5/4} \, (e\pi)^{(r-2)/4} \, R$$

where R is a numeric expression of integers. The factor of e appears to have an origin other than geometry. It is possibly a result of Quantum Perturbation Theory. The appearance of the $e\pi$ factor is interesting in light of its appearance in expressions for constants in chapter 3.

5. A View of the Coupling Constant Sequence

In this section we give an inverted view of the sequence of values of SU(n) coupling constants showing the correspondence with the Cosmos Theory sequence of spaces. It is clearly analogous to the Limos sector sequence of spaces. See Fig. 5.1. It differs in having a minimum for U(0).[14] The cause of the quadrupling sequence appears to be simple a counting of factors representing the number of fermions for the fundamental representation of each symmetry group. See section 4.2.

The below chart compares coupling constants with analogous Limos dimension array sizes. The sequence of coupling constant values is comparable to dimension array sizes.

Interaction	n	Coupling Constant Expression	Value	r	d_{dr}
SU(∞)		$e^2/0$		$-\infty$	∞
		.			
		.			
		.			
SU(7)		$e^2/\frac{1}{4}$		-2	4
SU(6)		$e^2/1$		-4	1
SU(5)		$e^2/4$		-6	$\frac{1}{4}$
SU(4)[15] ?		$e^2/16$	0.462	-8	1/16
SU(3)		$e^2/64$	0.115	-10	1/64
SU(2)		$e^2/256$	0.0289	-12	1/256
U(1) α		$e^2/1024$	0.00721438	-14	1/1024
U(0)		$e^2/4096$	0.0018036	-16	1/4096

5.1 Dimensions and Coupling Constants are "Equivalent"

Cosmos Theory spaces and particle interaction strengths (and thus interactions) all follow from geometry! No interactions, then no spaces with dynamics. No spaces, then no interactions.

Thus we find regularities in space and interaction constants ranging from Cosmos Theory to universes to universe properties to elementary particle interactions.

The appearance of such simple expressions for fundamental quantities suggests that there is a deeper geometric level to Cosmos Theory and Reality.

---

[14] Unless one considers fractional symmetry interactions U(1/n) such as we have in earlier books on gambols. Since gambols were of interest in particle studies where interactions are also of interest, the appearance of a Limos-like sequence is provocative.

[15] This value is based on the "doubling trend" seen in the three known coupling constants above.

COSMOS SPACES SPECTRUM

Blaha Space Number $N = o_s$	Cayley-Dickson Number n	Cayley Number d_c	Dimension Array column length d_{cn}	Dimension Array Size d_{dn}	Space-time-Dimension r
0	10	1024	2048	2048^2	18
1	9	512	1024	1024^2	16
2	8	256	512	512^2	14
3	7	128	256	256^2	12
4	6	64	128	128^2	10
5	5	32	64	64^2	8
6	4	16	32	32^2	6
7	**3**	**8**	**16**	**16^2**	**4**
8	2	4	8	8^2	2
9	1	2	4	4^2	0
10	0	1	2	2^2	-2
Limos :					
11	-1	½	1	1	-4
12	-2	¼	½	$½^2$	-6
13	-3	1/8	¼	$¼^2$	-8
14	-4	1/16	1/8	$1/8^2$	-10

•
•
•

HYPERCOSMOS OF THE SECOND KIND SPACES SPECTRUM

Blaha Space Number $N = O_s$	Cayley-Dickson Number n	Cayley Number d_c	Dimension Array size d_{dN2}	Space-time-Dimension r	CASe Group $su(2^{r/2},2^{r/2})$ CASe
0	10	1024	1024×2048	18	su(512,512)
1	9	512	512×1024	16	su(256,256)
2	8	256	256×512	14	su(128,128)
3	7	128	128×256	12	su(64,64)
4	6	64	64×128	10	su(32,32)
5	5	32	32×64	8	su(16,16)
6	4	16	16×32	6	su(8,8)
7	**3**	**8**	**8×16**	**4**	**su(4,4)**
8	**2**	4	4×8	2	su(2,2)
9	**1**	2	2×4	0	su(1,1)
10	0	1	1×2	-2	
11	-2	½	½	-4	

Figure 5.1. The HyperCosmos, Limos, and the HyperCosmos of the Second Kind, space spectrums.

REFERENCES

Akhiezer, N. I., Frink, A. H. (tr), 1962, *The Calculus of Variations* (Blaisdell Publishing, New York, 1962).

Bjorken, J. D., Drell, S. D., 1964, *Relativistic Quantum Mechanics* (McGraw-Hill, New York, 1965).

Bjorken, J. D., Drell, S. D., 1965, *Relativistic Quantum Fields* (McGraw-Hill, New York, 1965).

Blaha, S., 1995, *C++ for Professional Programming* (International Thomson Publishing, Boston, 1995).

________, 1998, *Cosmos and Consciousness* (Pingree-Hill Publishing, Auburn, NH, 1998 and 2002).

________, 2002, *A Finite Unified Quantum Field Theory of the Elementary Particle Standard Model and Quantum Gravity Based on New Quantum Dimensions™ & a New Paradigm in the Calculus of Variations* (Pingree-Hill Publishing, Auburn, NH, 2002).

________, 2004, *Quantum Big Bang Cosmology: Complex Space-time General Relativity, Quantum Coordinates™ Dodecahedral Universe, Inflation, and New Spin 0, ½, 1 & 2 Tachyons & Imagyons* (Pingree-Hill Publishing, Auburn, NH, 2004).

________, 2005a, *Quantum Theory of the Third Kind: A New Type of Divergence-free Quantum Field Theory Supporting a Unified Standard Model of Elementary Particles and Quantum Gravity based on a New Method in the Calculus of Variations* (Pingree-Hill Publishing, Auburn, NH, 2005).

________, 2005b, *The Metatheory of Physics Theories, and the Theory of Everything as a Quantum Computer Language* (Pingree-Hill Publishing, Auburn, NH, 2005).

________, 2005c, *The Equivalence of Elementary Particle Theories and Computer Languages: Quantum Computers, Turing Machines, Standard Model, Superstring Theory, and a Proof that Gödel's Theorem Implies Nature Must Be Quantum* (Pingree-Hill Publishing, Auburn, NH, 2005).

________, 2006a, *The Foundation of the Forces of Nature* (Pingree-Hill Publishing, Auburn, NH, 2006).

________, 2006b, *A Derivation of ElectroWeak Theory based on an Extension of Special Relativity; Black Hole Tachyons; & Tachyons of Any Spin.* (Pingree-Hill Publishing, Auburn, NH, 2006).

________, 2007a, *Physics Beyond the Light Barrier: The Source of Parity Violation, Tachyons, and A Derivation of Standard Model Features* (Pingree-Hill Publishing, Auburn, NH, 2007).

________, 2007b, *The Origin of the Standard Model: The Genesis of Four Quark and Lepton Species, Parity Violation, the ElectroWeak Sector, Color SU(3), Three Visible Generations of Fermions, and One Generation of Dark Matter with Dark Energy* (Pingree-Hill Publishing, Auburn, NH, 2007).

________, 2008a, *A Direct Derivation of the Form of the Standard Model From GL(16) (Pingree-Hill Publishing, Auburn, NH, 2008).*

________, 2008b, *A Complete Derivation of the Form of the Standard Model With a New Method to Generate Particle Masses Second Edition* (Pingree-Hill Publishing, Auburn, NH, 2008)

________, 2009, *The Algebra of Thought & Reality: The Mathematical Basis for Plato's Theory of Ideas, and Reality Extended to Include A Priori Observers and Space-Time Second Edition* (Pingree-Hill Publishing, Auburn, NH, 2009).

______, 2010a, *Operator Metaphysics: A New Metaphysics Based on a New Operator Logic and a New Quantum Operator Logic that Lead to a Mathematical Basis for Plato's Theory of Ideas and Reality* (Pingree-Hill Publishing, Auburn, NH, 2010).

______, 2010b, *The Standard Model's Form Derived from Operator Logic, Superluminal Transformations and GL(16)* (Pingree-Hill Publishing, Auburn, NH, 2010).

______, 2010c, *SuperCivilizations: Civilizations as Superorganisms* (McMann-Fisher Publishing, Auburn, NH, 2010).

______, 2011a, *21st Century Natural Philosophy Of Ultimate Physical Reality* (McMann-Fisher Publishing, Auburn, NH, 2011).

______, 2011b, *All the Universe! Faster Than Light Tachyon Quark Starships & Particle Accelerators with the LHC as a Prototype Starship Drive Scientific Edition* (Pingree-Hill Publishing, Auburn, NH, 2011).

______, 2011c, *From Asynchronous Logic to The Standard Model to Superflight to the Stars* (Blaha Research, Auburn, NH, 2011).

______, 2012a, *From Asynchronous Logic to The Standard Model to Superflight to the Stars volume 2: Superluminal CP and CPT, U(4) Complex General Relativity and The Standard Model, Complex Vierbein General Relativity, Kinetic Theory, Thermodynamics* (Blaha Research, Auburn, NH, 2012).

______, 2012b, *Standard Model Symmetries, And Four And Sixteen Dimension Complex Relativity; The Origin Of Higgs Mass Terms* (Blaha Reasearch, Auburn, NH, 2012).

______, 2013a, *Multi-Stage Space Guns, Micro-Pulse Nuclear Rockets, and Faster-Than-Light Quark-Gluon Ion Drive Starships* (Blaha Research, Auburn, NH, 2013).

______, 2013b, *The Bridge to Dark Matter; A New Sibling Universe; Dark Energy; Inflatons; Quantum Big Bang; Superluminal Physics; An Extended Standard Model Based on Geometry* (Blaha Reasearch, Auburn, NH, 2013).

______, 2014a, *Universes and Megaverses: From a New Standard Model to a Physical Megaverse; The Big Bang; Our Sibling Universe's Wormhole; Origin of the Cosmological Constant, Spatial Asymmetry of the Universe, and its Web of Galaxies; A Baryonic Field between Universes and Particles; Megaverse Extended Wheeler-DeWitt Equation* (Blaha Reasearch, Auburn, NH, 2014).

______, 2014b, *All the Megaverse! Starships Exploring the Endless Universes of the Cosmos Using the Baryonic Force* (Blaha Research, Auburn, NH, 2014).

______, 2014c, *All the Megaverse! II Between Megaverse Universes: Quantum Entanglement Explained by the Megaverse Coherent Baryonic Radiation Devices – PHASERs Neutron Star Megaverse Slingshot Dynamics Spiritual and UFO Events, and the Megaverse Microscopic Entry into the Megaverse* (Blaha Research, Auburn, NH, 2014).

______, 2015a, *PHYSICS IS LOGIC PAINTED ON THE VOID: Origin of Bare Masses and The Standard Model in Logic, U(4) Origin of the Generations, Normal and Dark Baryonic Forces, Dark Matter, Dark Energy, The Big Bang, Complex General Relativity, A Megaverse of Universe Particles* (Blaha Research, Auburn, NH, 2015).

______, 2015b, *PHYSICS IS LOGIC Part II: The Theory of Everything, The Megaverse Theory of Everything, U(4)$\otimes$U(4) Grand Unified Theory (GUT), Inertial Mass = Gravitational Mass, Unified Extended Standard Model and a New Complex General Relativity with Higgs Particles, Generation Group Higgs Particles* (Blaha Research, Auburn, NH, 2015).

______, 2015c, *The Origin of Higgs ("God") Particles and the Higgs Mechanism: Physics is Logic III, Beyond Higgs – A Revamped Theory With a Local Arrow of Time, The Theory of Everything Enhanced, Why Inertial Frames are Special, Universes of the Mind* (Blaha Research, Auburn, NH, 2015).

______, 2015d, *The Origin of the Eight Coupling Constants of The Theory of Everything: U(8) Grand Unified Theory of Everything (GUTE), S^8 Coupling Constant Symmetry, Space-Time Dependent Coupling Constants, Big Bang Vacuum Coupling Constants, Physics is Logic IV* (Blaha Research, Auburn, NH, 2015).

______, 2016a, *New Types of Dark Matter, Big Bang Equipartition, and A New U(4) Symmetry in the Theory of Everything: Equipartition Principle for Fermions, Matter is 83.33% Dark, Penetrating the Veil of the Big Bang, Explicit QFT Quark Confinement and Charmonium, Physics is Logic V* (Blaha Research, Auburn, NH, 2016).

______, 2016b, *The Periodic Table of the 192 Quarks and Leptons in The Theory of Everything: The U(4) Layer Group, Physics is Logic VI* (Blaha Research, Auburn, NH, 2016).

______, 2016c, *New Boson Quantum Field Theory, Dark Matter Dynamics, Dark Matter Fermion Layer Mixing, Genesis of Higgs Particles, New Layer Higgs Masses, Higgs Coupling Constants, Non-Abelian Higgs Gauge Fields, Physics is Logic VII* (Blaha Research, Auburn, NH, 2016).

______, 2016d, *Unification of the Strong Interactions and Gravitation: Quark Confinement Linked to Modified Short-Distance Gravity; Physics is Logic VIII* (Blaha Research, Auburn, NH, 2016).

______, 2016e, *MoND: Unification of the Strong Interactions and Gravitation II, Quark Confinement Linked to Large-Scale Gravity, Physics is Logic IX* (Blaha Research, Auburn, NH, 2016).

______, 2016f, *CQ Mechanics: A Unification of Quantum & Classical Mechanics, Quantum/Semi-Classical Entanglement, Quantum/Classical Path Integrals, Quantum/Classical Chaos* (Blaha Research, Auburn, NH, 2016).

______, 2016g, *GEMS Unified Gravity, ElectroMagnetic and Strong Interactions: Manifest Quark Confinement, A Solution for the Proton Spin Puzzle, Modified Gravity on the Galactic Scale* (Pingree Hill Publishing, Auburn, NH, 2016).

______, 2016h, *Unification of the Seven Boson Interactions based on the Riemann-Christoffel Curvature Tensor* (Pingree Hill Publishing, Auburn, NH, 2016).

______, 2017a, *Unification of the Eleven Boson Interactions based on 'Rotations of Interactions'* (Pingree Hill Publishing, Auburn, NH, 2017).

______, 2017b, *The Origin of Fermions and Bosons, and Their Unification* (Pingree Hill Publishing, Auburn, NH, 2017).

______, 2017c, *Megaverse: The Universe of Universes* (Pingree Hill Publishing, Auburn, NH, 2017).

______, 2017d, *SuperSymmetry and the Unified SuperStandard Model* (Pingree Hill Publishing, Auburn, NH, 2017).

______, 2017e, *From Qubits to the Unified SuperStandard Model with Embedded SuperStrings: A Derivation* (Pingree Hill Publishing, Auburn, NH, 2017).

______, 2017f, *The Unified SuperStandard Model in Our Universe and the Megaverse: Quarks, … ,* (Pingree Hill Publishing, Auburn, NH, 2017).

______, 2018a, *The Unified SuperStandard Model and the Megaverse SECOND EDITION A Deeper Theory based on a New Particle Functional Space that Explicates Quantum Entanglement Spookiness (Volume 1)* (Pingree Hill Publishing, Auburn, NH, 2018).

_______, 2018b, *Cosmos Creation: The Unified SuperStandard Model, Volume 2, SECOND EDITION* (Pingree Hill Publishing, Auburn, NH, 2018).

_______, 2018c, *God Theory (*Pingree Hill Publishing, Auburn, NH, 2018).

_______, 2018d, *Immortal Eye: God Theory: Second Edition* (Pingree Hill Publishing, Auburn, NH, 2018).

_______, 2018e, *Unification of God Theory and Unified SuperStandard Model THIRD EDITION* (Pingree Hill Publishing, Auburn, NH, 2018).

_______, 2019a, *Calculation of: QED α = 1/137, and Other Coupling Constants of the Unified SuperStandard Theory* (Pingree Hill Publishing, Auburn, NH, 2019).

_______, 2019b, *Coupling Constants of the Unified SuperStandard Theory SECOND EDITION* (Pingree Hill Publishing, Auburn, NH, 2019).

_______, 2019c, *New Hybrid Quantum Big_Bang–Megaverse_Driven Universe with a Finite Big Bang and an Increasing Hubble Constant* (Pingree Hill Publishing, Auburn, NH, 2019).

_______, 2019d, *The Universe, The Electron and The Vacuum* (Pingree Hill Publishing, Auburn, NH, 2019).

_______, 2019e, *Quantum Big Bang – Quantum Vacuum Universes (Particles)* (Pingree Hill Publishing, Auburn, NH, 2019).

_______, 2019f, *The Exact QED Calculation of the Fine Structure Constant Implies ALL 4D Universes have the Same Physics/Life Prospects* (Pingree Hill Publishing, Auburn, NH, 2019).

_______, 2019g, *Unified SuperStandard Theory and the SuperUniverse Model: The Foundation of Science* (Pingree Hill Publishing, Auburn, NH, 2019).

_______, 2020a, *Quaternion Unified SuperStandard Theory (The QUeST) and Megaverse Octonion SuperStandard Theory (MOST)* (Pingree Hill Publishing, Auburn, NH, 2020).

_______, 2020b, *United Universes Quaternion Universe - Octonion Megaverse* (Pingree Hill Publishing, Auburn, NH, 2020).

_______, 2020c, *Unified SuperStandard Theories for Quaternion Universes & The Octonion Megaverse* (Pingree Hill Publishing, Auburn, NH, 2020).

_______, 2020d, *The Essence of Eternity: Quaternion & Octonion SuperStandard Theories* (Pingree Hill Publishing, Auburn, NH, 2020).

_______, 2020e, *The Essence of Eternity II* (Pingree Hill Publishing, Auburn, NH, 2020).

_______, 2020f, *A Very Conscious Universe* (Pingree Hill Publishing, Auburn, NH, 2020).

_______, 2020g, *Hypercomplex Universe* (Pingree Hill Publishing, Auburn, NH, 2020).

_______, 2020h, *Beneath the Quaternion Universe* (Pingree Hill Publishing, Auburn, NH, 2020).

_______, 2020i, *Why is the Universe Real? From Quaternion & Octonion to Real Coordinates* (Pingree Hill Publishing, Auburn, NH, 2020).

_______, 2020j, *The Origin of Universes: of Quaternion Unified SuperStandard Theory (QUeST); and of the Octonion Megaverse (UTMOST)* (Pingree Hill Publishing, Auburn, NH, 2020).

_______, 2020k, *The Seven Spaces of Creation: Octonion Cosmology* (Pingree Hill Publishing, Auburn, NH, 2020).

______, 2020l, *From Octonion Cosmology to the Unified SuperStandard Theory of Particles* (Pingree Hill Publishing, Auburn, NH, 2020).

______, 2021a, *Pioneering the Cosmos* (Pingree Hill Publishing, Auburn, NH, 2021).

______, 2021b, *Pioneering the Cosmos II* (Pingree Hill Publishing, Auburn, NH, 2021).

______, 2021c, *Beyond Octonion Cosmology* (Pingree Hill Publishing, Auburn, NH, 2021).

______, 2021d, *Universes are Particles* (Pingree Hill Publishing, Auburn, NH, 2021).

______, 2021e, *Octonion-like dna-based life, Universe expansion is decay, Emerging New Physics* (Pingree Hill Publishing, Auburn, NH, 2021).

______, 2021f, *The Science of Creation New Quantum Field Theory of Spaces* (Pingree Hill Publishing, Auburn, NH, 2021).

______, 2021g, *Quantum Space Theory With Application to Octonion Cosmology & Possibly To Fermionic Condensed Matter* (Pingree Hill Publishing, Auburn, NH, 2021).

______, 2021h, *21st Century Natural Philosophy of Octonion Cosmology , and Predestination, Fate, and Free Will* (Pingree Hill Publishing, Auburn, NH, 2021).

______, 2021i, *Beyond Octonion Cosmology II : Origin of the Quantum; A New Generalized Field Theory (GiFT); A Proof of the Spectrum of Universes; Atoms in Higher Universes* (Pingree Hill Publishing, Auburn, NH, 2021).

______, 2021j, *Integration of General Relativity and Quantum Theory: Octonion Cosmology, GiFT, Creation/Annihilation Spaces CASe, Reduction of Spaces to a Few Fermions and Symmetries in Fundamental Frames* (Pingree Hill Publishing, Auburn, NH, 2021).

______, 2022a, *New View of Octonion Cosmology Based on the Unification of General Relativity and Quantum Theory* (Pingree Hill Publishing, Auburn, NH, 2022).

______, 2022b, *The Dust Beneath Hypercomplex Cosmology* (Pingree Hill Publishing, Auburn, NH, 2022).

______, 2022c, *Passing Through Nature to Eternity: ProtoCosmos, HyperCosmos, Unified SuperStandard Theory* (Pingree Hill Publishing, Auburn, NH, 2022).

______, 2022d, *HyperCosmos Fractionation and Fundamental Reference Frame Based Unification: Particle Inner Space Basis of Parton and Dual Resonance Models* (Pingree Hill Publishing, Auburn, NH, 2022).

______, 2022e, *A New UniDimension ProtoCosmos and SuperString F-Theory Relation to the HyperCosmos* (Pingree Hill Publishing, Auburn, NH, 2022).

______, 2022f, *The Cosmic Panorama: ProtoCosmos, HyperCosmos,Unified SuperStandard Theory (UST) Derivation* (Pingree Hill Publishing, Auburn, NH, 2022).

______, 2022g, *Ultimate Origin: ProtoCosmos and HyperCosmos* (Pingree Hill Publishing, Auburn, NH, 2022).

______, 2023a, *UltraUnification and the Generation of the Cosmos* (Pingree Hill Publishing, Auburn, NH, 2023).

______, 2023b, *God and and Cosmos Theory* (Pingree Hill Publishing, Auburn, NH, 2023).

______, 2023c, *A New Completely Geometric SU(8) Cosmos Theory; New PseudoFermion Fields; Fibonacci-like Dimension Arrays; Ramsey Number Approximation* (Pingree Hill Publishing, Auburn, NH, 2023).

______, 2023d, *Newton's Apple is Now the Fermion* (Pingree Hill Publishing, Auburn, NH, 2023).

_______, 2023e,*Cosmos Theory: The Sub-Particle Gambol Model* (Pingree Hill Publishing, Auburn, NH, 2023).

_______, 2024a, *Cosmos-Universe-Particle-Gambol Theory* (Pingree Hill Publishing, Auburn, NH, 2024).

_______, 2024b, *Fractal Cosmos Theory* (Pingree Hill Publishing, Auburn, NH, 2024).

_______, 2024c, *Fractal Cosmic Curve: Tensor-Based CosmosTheory* (Pingree Hill Publishing, Auburn, NH, 2024).

_______, 2024d, *The Eternal Form of Cosmos Theory* (Pingree Hill Publishing, Auburn, NH, 2024).

_______, 2024e, *The Eternal Form of Cosmos Theory Third Edition* (Pingree Hill Publishing, Auburn, NH, 2024).

Eddington, A. S., 1952, *The Mathematical Theory of Relativity* (Cambridge University Press, Cambridge, U.K., 1952).

Fant, Karl M., 2005, *Logically Determined Design: Clockless System Design With NULL Convention Logic* (John Wiley and Sons, Hoboken, NJ, 2005).

Feinberg, G. and Shapiro, R., 1980, *Life Beyond Earth: The Intelligent Earthlings Guide to Life in the Universe* (William Morrow and Company, New York, 1980).

Gelfand, I. M., Fomin, S. V., Silverman, R. A. (tr), 2000, *Calculus of Variations* (Dover Publications, Mineola, NY, 2000).

Giaquinta, M., Modica, G., Souchek, J., 1998, *Cartesian Coordinates in the Calculus of Variations* Volumes I and II (Springer-Verlag, New York, 1998).

Giaquinta, M., Hildebrandt, S., 1996, *Calculus of Variations* Volumes I and II (Springer-Verlag, New York, 1996).

Gradshteyn, I. S. and Ryzhik, I. M., 1965, *Table of Integrals, Series, and Products* (Academic Press, New York, 1965).

Heitler, W., 1954, *The Quantum Theory of Radiation* (Claendon Press, Oxford, UK, 1954).

Huang, Kerson, 1992, *Quarks, Leptons & Gauge Fields 2nd Edition* (World Scientific Publishing Company, Singapore, 1992).

Jost, J., Li-Jost, X., 1998, *Calculus of Variations* (Cambridge University Press, New York, 1998).

Kaku, Michio, 1993, *Quantum Field Theory*, (Oxford University Press, New York, 1993).

Kirk, G. S. and Raven, J. E., 1962, *The Presocratic Philosophers* (Cambridge University Press, New York, 1962).

Landau, L. D. and Lifshitz, E. M., 1987, *Fluid Mechanics 2nd Edition*, (Pergamon Press, Elmsford, NY, 1987).

Rescher, N., 1967, *The Philosophy of Leibniz* (Prentice-Hall, Englewood Cliffs, NJ, 1967).

Riesz, Frigyes and Sz.-Nagy, Béla, 1990, *Functional Analysis* (Dover Publications, New York, 1990).

Sakurai, J. J., 1964, *Invariance Principles and Elementary Particles* (Princeton University Press, Princeton, NJ, 1964).

Weinberg, S., 1972, *Gravitation and Cosmology* (John Wiley and Sons, New York, 1972).

Weinberg, S., 1995, *The Quantum Theory of Fields Volume I* (Cambridge University Press, New York, 1995).

INDEX

About the Author

Stephen Blaha is a well-known Physicist and Man of Letters with interests in Science, Society and civilization, the Arts, and Technology. He had an Alfred P. Sloan Foundation scholarship in college. He received his Ph.D. in Physics from Rockefeller University. He has served on the faculties of several major universities. He was also a Member of the Technical Staff at Bell Laboratories, a manager at the Boston Globe Newspaper, a Director at Wang Laboratories, and President of Blaha Software Inc. and of Janus Associates Inc. (NH).

Among other achievements he was a co-discoverer of the "r potential" for heavy quark binding developing the first (and still the only demonstrable) non-Aeolian gauge theory with an "r" potential; first suggested the existence of topological structures in superfluid He-3; first proposed Yang-Mills theories would appear in condensed matter phenomena with non-scalar order parameters; first developed a grammar-based formalism for quantum computers and applied it to elementary particle theories; first developed a new form of quantum field theory without divergences (thus solving a major 60 year old problem that enabled a unified theory of the Standard Model and Quantum Gravity without divergences to be developed); first developed a formulation of complex General Relativity based on analytic continuation from real space-time; first developed a generalized non-homogeneous Robertson-Walker metric that enabled a quantum theory of the Big Bang to be developed without singularities at t = 0; first generalized Cauchy's theorem and Gauss' theorem to complex, curved multi-dimensional spaces; received Honorable Mention in the Gravity Research Foundation Essay Competition in 1978; first developed a physically acceptable theory of faster-than-light particles; first derived a composition of extremums method in the Calculus of Variations; first quantitatively suggested that inflationary periods in the history of the universe were not needed; first proved Gödel's Theorem implies Nature must be quantum; provided a new alternative to the Higgs Mechanism, and Higgs particles, to generate masses; first showed how to resolve logical paradoxes including Gödel's Undecidability Theorem by developing Operator Logic and Quantum Operator Logic; first developed a quantitative harmonic oscillator-like model of the life cycle, and interactions, of civilizations; first showed how equations describing superorganisms also apply to civilizations. A recent book shows his theory applies successfully to the past 14 years of history and to *new* archaeological data on Andean and Mayan civilizations as well as Early Anatolian and Egyptian civilizations.

He first developed an axiomatic derivation of the form of The Standard Model from geometry – space-time properties – The Unified SuperStandard Model. It unifies all the known forces of Nature. It also has a Dark Matter sector that includes a Dark ElectroWeak sector with Dark doublets and Dark gauge interactions. It uses quantum coordinates to remove infinities that crop up in most interacting quantum field theories and additionally to remove the infinities that appear in the Big Bang and generate inflationary growth of the universe. It shows gravity has a MOND-like form without sacrificing Newton's Laws. It relates the interactions of the MOND-like sector of gravity with the r-potential of Quark Confinement. The axioms of the theory lead to the question of their origin. We suggest in the preceding edition of this book it can be attributed to an entity with God-like properties. We explore these properties in "God Theory" and show they predict that the Cosmos exists forever although individual universes (or incarnations of our universe) "come and go." Several other important results emerge from God Theory such a functionally triune God. The Unified SuperStandard Theory has many other important parts described in the Current Edition of *The Unified SuperStandard Theory* and expanded in subsequent volumes.

Blaha has had a major impact on a succession of elementary particle theories: his Ph.D. thesis (1970), and papers, showed that quantum field theory calculations to all orders in ladder approximations could not give scaling deep inelastic electron-nucleon scattering. He later showed the eigenvalue equation for the fine structure constant α in Johnson-Baker-Willey QED had a zero at $\alpha = 1$ not 1/137 by solving the Schwinger-Dyson equations to all orders in an approximation that agreed with exact results to 4th order in α thus ending interest in this theory. In 1979 at Prof. Ken Johnson's (MIT) suggestion he calculated the proton-neutron mass difference in the MIT bag model and found the result had the wrong sign reducing interest in the bag model. These results all appear in Physical Review papers. In the 2000's he repeatedly pointed out the shortcomings of SuperString theory and showed that The Standard Model's form could be derived from space-time geometry by an extension of Lorentz transformations to faster than light transformations. This deeper space-time basis greatly increases the possibility that it is part of THE fundamental theory. Recently, Blaha showed that the Weak interactions differed significantly from the Strong, electromagnetic and gravitation interactions in important respects while these interactions had similar features, and suggested that ElectroWeak theory, which is essentially a glued union of the Weak interactions and Electromagnetism, possibly modulo unknown Higgs particle features, be replaced by a unified theory of the other interactions combined with a stand-alone Weak interaction theory. Blaha also showed that, if Charmonium calculations are taken seriously, the Strong interaction coupling constant is only a factor of five larger than the electromagnetic coupling constant, and thus Strong interaction perturbation theory would make sense and yield physically meaningful results.

In graduate school (1965-71) he wrote substantial papers in elementary particles and group theory: The Inelastic E- P Structure Functions in a Gluon Model. Phys. Lett. B40:501-502,1972; Deep-Inelastic E-P Structure Functions In A Ladder Model With Spin 1/2 Nucleons, Phys.Rev. D3:510-523,1971; Continuum Contributions To The Pion Radius, Phys. Rev. 178:2167-2169,1969; Character Analysis of U(N) and SU(N), J. Math. Phys. 10, 2156 (1969); and The Calculation of the Irreducible Characters of the Symmetric Group in Terms of the Compound Characters, (Published as Blaha's Lemma in D. E. Knuth's book: *The Art of Computer Programming Vols. 1 – 4*).

In the early 1980's Blaha was also a pioneer in the development of UNIX for financial, scientific and Internet applications: benchmarked UNIX versions showing that block size was critical for UNIX performance, developing financial modeling software, starting database benchmarking comparison studies, developing Internet-like UNIX networking (1982) and developing a hybrid shell programming technique (1982) that was a precursor to the PERL programming language. He was also the manager of the AT&T ten-year future products development database. His work helped lead to commercial UNIX on computers such as Sun Micros, IBM AIX minis, and Apple computers.

In the 1980's he pioneered the development of PC Desktop Publishing on laser printers and was nominated for three "Awards for Technical Excellence" in 1987 by PC Magazine for PC software products that he designed and developed.

Recently he has developed a theory of Megaverses – actual universes of which our universe is one – with quantum particle-like properties based on the Wheeler-DeWitt equation of Quantum Gravity. He has developed a theory of a baryonic force, which had been conjectured many years ago, and estimated the strength of the force based on discrepancies in measurements of the gravitational constant G. This force, operative in D-dimensional space, can be used to escape from our universe in "uniships" which are the equivalent of the faster-than-light starships proposed in the author's earlier books. Thus travel to other universes, as well as to other stars is possible.

Blaha also considered the complexified Wheeler-DeWitt equation and showed that its limitation to real-valued coordinates and metrics generated a Cosmological Constant in the Einstein equations.

The author has also recently written a series of books on the serious problems of the United States and their solution as well as a book on the decline of Mankind that will follow from current social and genetic trends in Mankind.

In the past twenty years Dr. Blaha has written over 80 books on a wide range of topics. Some recent major works are: *From Asynchronous Logic to The Standard Model to Superflight to the Stars, All the Universe!, SuperCivilizations: Civilizations as Superorganisms, America's Future: an Islamic Surge, ISIS, al Qaeda, World Epidemics, Ukraine, Russia-China Pact, US Leadership Crisis, The Rises and Falls of Man – Destiny – 3000 AD: New Support for a Superorganism MACRO-THEORY of CIVILIZATIONS From CURRENT WORLD TRENDS and NEW Peruvian, Pre-Mayan, Mayan, Anatolian, and Early Egyptian Data, with a Projection to 3000 AD,* and *Mankind in Decline: Genetic Disasters, Human-Animal Hybrids, Overpopulation, Pollution, Global Warming, Food and Water Shortages, Desertification, Poverty, Rising Violence, Genocide, Epidemics, Wars, Leadership Failure.*

He has taught approximately 4,000 students in undergraduate, graduate, and postgraduate corporate education courses primarily in major universities, and large companies and government agencies.

He developed a quantum theory, The Unified SuperStandard Theory (UST), which describes elementary particles in detail without the difficulties of conventional quantum field theory. He found that the internal symmetries of this theory could be exactly derived from an octonion theory called QUeST. He further found that another octonion theory (UTMOST) describes the Megaverse. It can hold QUeST universes such as our own universe. It has an internal symmetry structure which is a superset of the QUeST internal symmetries.

Recently he developed Octonion Cosmology. He replaced it with HyperCosmos theory, which has significantly better features. He developed a fractionalization process for dimensions, particles and symmetry groups. He also described transformation that reduced particles and dimensions to a far more compact form. He also developed a precursor theory ProtoCosmos that leads to the HyperCosmos.

The author showed that space-time and Internal Symmetries can be unified in any of the ten HyperCosmos spaces in their associated HyperUnification spaces. The combined set of HyperUnification spaces enable all HyperCosmos dimensions to be obtained by a General Relativistic transformation from one primordial dimension in the 42 space-time dimension unified HyperUnification space.

At present the author devel;oped the Cosmos Theory that incorporates ProtoCosmos Theory, HyperCosmos Theory, Limos Theory, Second Kind HyperCosmos Theory and HyperUnification Spaces. He has introduced PseudoFermion wave functions and theory, He has related Cosmos Theory to Regge trajectories of spaces, parton theory, Veneziano amplitudes, Fibonacci numbers and Ramsey numbers. He has calculated an approximation to the difficult R(n,n) Ramsey numbers.

He has developed a Gambol Model that successfully accounts for e-p deep inelastic scattering, fundamental particle resonances, hadron scattering, and the inner structure of particles based on confinement through Casimir forces of ideal gambol gases. The Gambol Planckian Distribution was derived.

He has applied the Gambol Model to particles, universes, and the Cosmos of universes. He showed that the Cosmos may have a distribution of 23 universes corresponding to various Cosmos spaces.

Recently he showed that Cosmos Theory follows from the number of independent asymmetric tensors in a dimension r. He also showed the close parallel between the form of γ-matrices and Cosmos Theory dimension arrays. The closeness suggested that dimension arrays have the same importance as γ-matrices for fermions.

He demonstrated that the pressure of fermions within a space of dimension r balances the Casimir vacuum energy force for 18 dimensions. He showed that $2e\pi = 17.02$ marks the critical point where pressure balances Casimir force, which implies $r = 18$ is the highest dimension Physical Cosmos space. The dimension $2e\pi$ appears to set the approximate dimension for Cosmos spaces with dimension array size $2^{r+4} \cong (17.02/8)^{r+4}. \cong (e\pi/4)^{r+4}. \cong 2.13^{r+4}.$

Now he has found the sequence of Coupling Constant values in the Standard Model and UST.